MW01264330

SKILLED AND VOCATIONAL TRADES

BECOME AN IT SUPPORT TECHNICIAN

by Sheryl Normandeau

BrightPoint Press

San Diego, CA

BrightPoint Press

For more information, contact:
BrightPoint Press
PO Box 27779
San Diego, CA 92198
www.BrightPointPress.com

LIBRARY OF CONGRESS CATALOGING-IN-PUBLICATION DATA

Names: Normandeau, Sheryl, author.
Title: Become an IT support technician / by Sheryl Normandeau.
Description: San Diego, CA : BrightPoint Press, [2023] | Series: Skilled and vocational trades | Includes bibliographical references and index. | Audience: Grades 10-12
Identifiers: ISBN 9781678204181 (hardcover) | ISBN 9781678204198 (eBook)
The complete Library of Congress record is available at www.loc.gov

CONTENTS

AT A GLANCE

- Information technology (IT) support technicians keep computer systems and other related technology in working order for clients and coworkers.
- IT support technicians have roles in large corporations, governments, charitable organizations, retail, industry, manufacturing, health care, schools, libraries, and banks.
- Some IT support technicians don't work for a specific company. They may be freelancers who work on their own.
- IT support technicians may work alone. They may work as part of a team. It all depends on the type of work they do.
- People skills are a huge part of an IT support technician's job.

- Someone who is interested in working as an IT support technician must obtain a high school diploma or pass a General Educational Development (GED) test. For many higher-tier IT support technician jobs, a university degree is required.
- In 2020, according to the US Bureau of Labor Statistics, the need for IT support technicians was expected to increase.
- As sectors such as telecommunications and health care rely more on IT to help serve their clients, the need for IT support technicians will grow.

A DAY IN THE LIFE

Zach Hill is an information technology (IT) career coach. He formerly worked for a large school district. He and his team of IT support technicians made sure that each school's computers and devices were working. They supported the work of school **administrators**, office staff, teachers, and students.

IT support technicians help students in computer labs.

Hill and his team fixed printers and tablets. They also set up computer labs and hardware such as smart whiteboards. Approximately 9,000 people relied on Hill

IT support technicians team up to plan big projects.

and his team to keep their computers and other devices working. Hill says, “One of the things you learn if you’re going into IT support for education is if it plugs in, you’ll find that you end up supporting it in some way.”[1]

Hill's career as a technician required him to be flexible. Often he could **diagnose** problems and provide solutions **remotely**. Other times he had to travel to the school. Sometimes he and another IT support technician collaborated on a project. Other times he worked on his own. When a large project was due, the whole team worked together.

SUPPORTING AN ONLINE WORLD

Information technology is the use of computers, other devices, and the internet to get work done. Nearly any organization that uses computers likely needs technicians. Technicians have roles in governments, charities, and schools. They keep hospitals and banks operating efficiently. They also work in the retail and manufacturing industries.

Some technicians don't work for a specific company. They may be freelancers who work on their own. They are hired to work on short-term projects. They might install software for a new business, for example. Once they finish that job, they move on to the next one. Still other technicians work at help desks or call centers. They give customers

Dozens of IT support technicians may work in a single large call center.

solutions to problems. They help with with computers, telephones, and other related devices.

Experts expect the need for technicians to increase. Modern companies rely on computers and networks more than

The need for IT support technicians will continue to grow as companies rely more on computers.

ever before. Industries are adapting to serve people online. Governments and hospitals provide services virtually. Technicians will ensure this technology keeps running smoothly.

CHAPTER ONE

WHAT DOES AN IT SUPPORT TECHNICIAN DO?

IT support technicians keep computer systems and other related technology in working order for clients and coworkers. Personal computers (PCs) were affordable enough for some in the 1970s. It wasn't until the 1990s that computers really became an

Older desktop computers were larger and slower than today's laptops.

important part of people's lives. Changes in technology made PCs faster and smaller. Laptops let people take powerful computers anywhere. And people started using the internet. The internet is a network that connects computers across the globe.

Modern computers help to keep businesses and governments running. Without them, much of daily life would be disrupted. They are used in banking and commerce. Computers are tools in schools. They can help save the lives of patients in a hospital. When a car needs repairs, a computer helps find the problem.

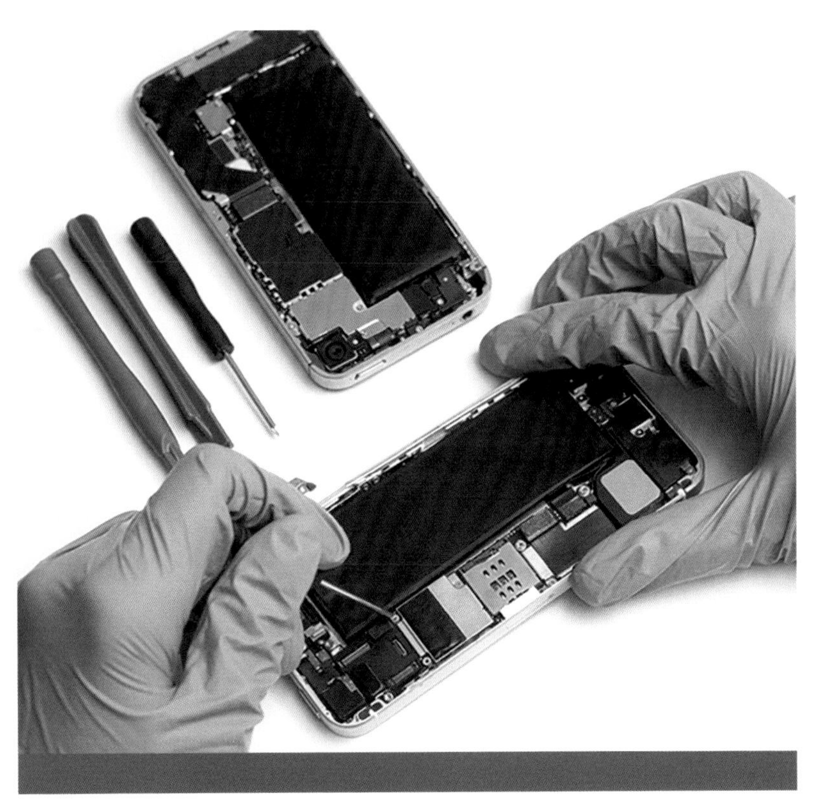

IT support technicians fix many devices aside from computers, including cell phones, tablets, and televisions.

Many devices contain computer technology. This includes cell phones. The first handheld cell phone was made in 1973. Today, 85 percent of Americans have smartphones. Computers, phones, and other devices may break down. They run into problems. That's where technicians come in. They figure out what is wrong with the device and fix it.

PROBLEM-SOLVING SKILLS

Hard skills are specific abilities that technicians need. Technicians use a variety of skills every day. One important hard skill is analyzing problems with computers, devices, and networks. These devices may stop working. Technicians must diagnose the problem. Then they need to find a solution.

A computer's hardware, or physical parts, may break down, requiring the IT support technician to fix it.

Mai Yia Vue works as a help desk specialist for a national bank based in Minnesota. She says, "It's a huge responsibility, as issues can range from software features not working correctly

Keeping records about devices is an important duty for IT support technicians. Past records can help solve future problems.

to whole servers being down." She offers advice to people interested in this career. "The learning process is ongoing. . . . Do your best and ask for help when needed."[2]

Technicians may also keep records about the devices. They will write down information about how long the devices have been in use. They will take notes on problems the devices have had in the past. Technicians will keep track of each time the device needs to be fixed. The will record how they fixed the problem. Knowing a device's past problems can help technicians solve current issues with it.

Help desk technicians may provide support to employees over the phone or with a headset.

PROVIDING SUPPORT

Technicians may work for a specific company at a help desk or call center. Customers will call them for help when computers or other devices break down. The help desk technicians look at the issue. They diagnose computer issues over the phone or in an online chat. They directly

ESCALATING A TECH SUPPORT CALL

If the customer cannot fix a problem with the support of the help desk, the call is escalated. That means the help desk technician asks a higher-level expert to solve the issue. The other technician may connect remotely to fix the computer. Sometimes, the technician will need to visit the customer to fix the problem.

connect with customers. Technicians do many other lesser-known tasks. They remove viruses, transfer computer data, and help employees log in to the company's systems.

Technicians also train other technicians. They may train new employees. They also often train existing employees if a company's technology is updated. Large companies may have a big team of technicians on staff. Smaller companies may hire only one or two technicians. Very small companies may use freelancers instead.

At a large company, IT support technicians may work with other employees from different departments.

WORKING STYLES

Depending on the type of job they do, technicians may work alone. Or they may work as part of a team. Freelance technicians are usually self-employed. They are their own bosses. They decide which contracts to take. They also choose who to work for.

Technicians who work for a company will have a supervisor who manages the team. The supervisor makes sure all the work is done properly and on time. In large companies, they may work with other types of employees. These workers might specialize in departments such as customer service, security, or human resources. Technicians must be able to work with people with different roles and skills.

CUSTOMER RELATIONS

Technicians may work closely with customers. This is especially true for help desk technicians. They get requests for help directly from the customers. Those who work in retail stores also have direct contact with customers.

CAREER MOVEMENT

There are four different tiers of IT support. Technicians can learn new skills and train to work at different levels. At the higher levels, the pay increases. So do the duties. Technicians in Tier 1, the lowest tier, help customers solve basic computer problems. Higher-tier technicians, like those in Tier 4, specialize in jobs outside of the organization. They may help with printers or software.

Spencer Wyrick is an IT support technician who works for a car dealership. Problem solving is always a part of Wyrick's work. He says, "There was a client . . . who had hearing aids, and he was having trouble using our phone system. So I was able to work with him and get a new phone for him that we could actually connect via Bluetooth to his hearing aids, and he was able to hear the client and talk to them." Wyrick explains that being able to help people is rewarding. "Just to see him happy and . . . thrilled to be able to work easier was just . . . a good feeling."[3]

Working in a retail store allows IT support technicians to use their communication skills while interacting directly with customers.

CHAPTER TWO

WHAT TRAINING DO IT SUPPORT TECHNICIANS NEED?

The type of training IT support technicians need depends on their job path. A help desk technician may not need the same education as someone in a higher tier of IT support. Education for technicians varies. People can earn **certifications** or

High school computer science classes teach students practical skills for a future in information technology.

college degrees. Some teach themselves. They can also learn on the job through an **apprenticeship**. More education usually leads to a better-paying position.

EDUCATION

People who are interested in working as technicians must get a high school diploma. Or they can instead pass a General Educational Development (GED) test. When choosing courses in high school, they should focus on science and math.

HELP DESKS IN HIGH SCHOOLS

Some schools encourage students to operate IT support help desks in their own schools. The students running the help desk troubleshoot any problems the other students are having with the technology used in the school. This is a practical way to prepare students for careers as IT support technicians. It gives students practice with communication and problem-solving.

If possible, they should take computer science classes.

A college degree is required for many higher-tier technician jobs. A bachelor's in computer science is a common choice. Studying information technology or computer information systems are other options.

CERTIFICATION

Becoming certified can open up more career options for technicians. To do this, they take special courses. They must complete coursework and pass exams to earn certification. Some certification

Certification programs offer more specific training and take less time than a college degree.

programs are run by technology industry leaders.

The Computing Technology Industry Association A+ (CompTIA A+) certification is a popular certification program in IT support. By taking this program, students learn about computer operating systems and security. They also learn how to store data and work at a help desk. Students must pass two tests to earn the certificate. CompTIA A+ is considered an entry-level (Tier 1) certification. Other certifications are available, as well.

APPRENTICESHIP

Some companies will hire people studying IT support. These companies will allow the students to work and learn on the job. This is called an apprenticeship. It can provide pay and education. For some students, this is a valuable opportunity.

Apprentices work under more experienced employees to gain on-the-job training.

Aaron Chidgey works as an IT support technician for a large construction company. He did his training and worked in IT at the same time. He says, "I got into the construction industry by completing a two-year IT apprenticeship. . . . I found that an apprenticeship was a great way to learn whilst on the job. Not only are you earning, but you're learning in a real-world environment."[4]

VOLUNTEERING IN IT SUPPORT

Volunteering to work with computers can help high school students learn skills for the future. It can be great on a resume. It shows that the student is reliable, responsible, and engaged in the community. In some cases, students can even earn credits toward a diploma by doing volunteer work. Nonprofits, churches, and community organizations often need help with computers. Volunteers complete tasks like setting up devices and developing websites.

LIFELONG LEARNING

The learning never stops when working as a technician. Technology changes quickly. Technicians must adapt to it. They need to be able to give the best customer service they can.

Clay is an IT support technician for Holland Hospital in Michigan. He talks about how advances in technology have changed his job. Clay says, “Technology is evolving to where the communication between employees, patients, and also other people within the hospital and organization need to better be able to meet

As technology changes, IT support technicians must learn new devices and programs to serve their customers.

the needs of those patients, and we can do that better by applying our technology that we have."[5] Technicians must keep up with their training, so that they can work with new technology that is introduced.

MENTORSHIP

Mentorship is important in the IT field. Mentors are people who have been working in the field for a long time. They are considered senior employees. Mentors have a lot of experience. They help new technicians feel confident in their jobs. Mentors teach new skills. They show practical, useful ways to do tasks.

Mai Yia Vue remarks on getting help as an IT support technician. As a Tier 1 specialist, she finds ways to learn from her team. “I always ask questions whenever I don’t understand something. I am lucky that

Mentors can help new IT support technicians gain opportunities and feel more comfortable in the workplace.

my service center always has a chat open with other technicians. We use this chat to post questions and help each other out," she says. "I reach out to my managers as well and they are always happy to help."[6]

Mentorship does not only happen between two employees. Many large companies have mentorship groups for employees, with a mix of senior and new employees.

There are many ways to mentor a new technician. A senior employee might ask the new staff member to follow him or her while he or she completes day-to-day tasks. This is called job shadowing. The mentor may encourage the new technician to participate in brainstorming ideas and problem solving with the rest of the team. This helps the new employee feel more comfortable with the job and other employees.

CHAPTER THREE

WHAT IS LIFE LIKE AS AN IT SUPPORT TECHNICIAN?

An IT support technician's career is fast-paced and interesting. These workers do a wide range of tasks. They communicate with many different people. Every day brings a unique set of problems to solve. They must be ready to help whenever someone is in need.

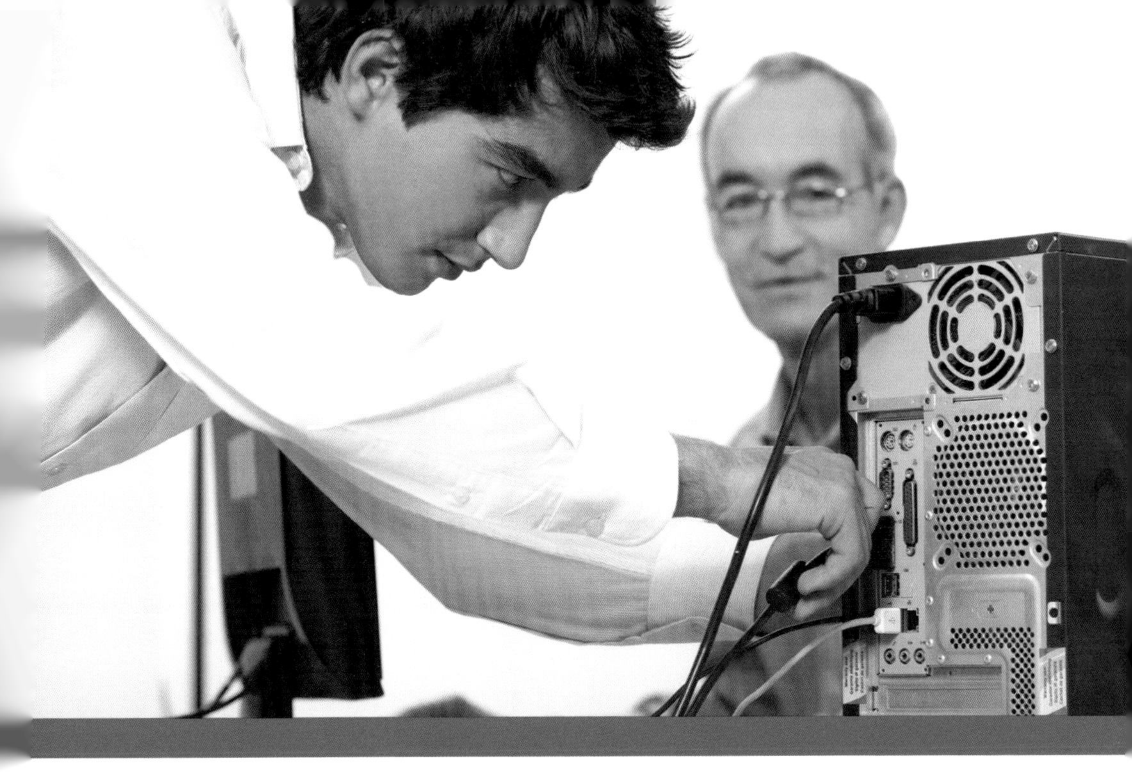

IT support technicians may fix computer monitors, keyboards, and speakers. They must understand computer hardware and software.

FREELANCING OR WORKING FOR AN EMPLOYER

Technicians who work as freelancers do not work only for one company. They receive contracts for work on various projects from different organizations. Their contract is **terminated** once they have finished working on the project. Then they start working on another one. Freelance technicians can pick and choose which types of contracts they work on. They can work as little or as much as they want. Freelance technicians have more control over the types of work they do. Technicians who work for a company usually have less. Work isn't always steady for a freelancer, however. They depend on clients for work. Some people prefer the security of regular work and pay.

If a technician is hired by a company, it usually means that there is a steady supply of work. The technician who works for an employer may not be able to choose the types of jobs they do. Even though the duties are assigned, they can still be interesting and varied.

WHAT IS TICKETING?

When a client needs help, they are said to "open a ticket" with the IT team. This creates a formal record of the problem. Clients can open a ticket in different ways. They may use a phone call, an online chat, or an email. The technician looks at the problem once the ticket is filed. He or she tries to solve it. After the problem is fixed, the ticket is closed.

HOURS AND BENEFITS

Technicians who are not freelancers may have a set schedule. They will work either part-time or full-time. Full-time work is generally forty hours per week. They usually have steady work. They may work overtime when there is a big project. Some IT roles are more structured than others. Christoph Puetz is an IT career coach. He explains, "Work in a call center is very structured, so you usually have fixed working hours, a fixed schedule."[7]

Some IT work involves long hours. Large crashes or problems take time

Set hours and steady work are benefits to working in a call center.

to fix. Freelance and contract projects may require extra time. Once the project has been completed, the freelance technician will look for other contracts. This can mean that there is a period when the technician is not working.

Sometimes technicians must work shifts or be on call. That means they must be available to work at any time of any day. This structure can create long hours and an irregular schedule. This may mean the technician may not spend as much time with family and friends. Dawn Fread is a former IT computer support specialist technician and a college instructor. She says, "For me, the challenge was raising children and being in IT. . . . You don't have typically normal 8 to 5 hours. . . . You would go in and you might have 12-hour days, and so it is juggling the family."[8] Not all

An on-call IT job may require more time responding to calls.

technician jobs require the technician to be on call, however.

There may be perks to working for a company. Some benefits include health insurance and retirement savings programs. Freelance technicians don't usually receive such benefits. This is something to consider when deciding whether to freelance or work for a company.

TOOLS OF THE TRADE

Technicians need soft skills. Soft skills include abilities like communication, time management, and working with people. Technicians also need to be able to prioritize their work. They must decide which tasks need to be done right away and which ones can wait. Meeting deadlines is an important duty of a technician. The technician has to consider the needs of the customer. It is important that the job be done quickly and properly.

Much of what a technician does requires critical thinking and other skills. Still, a few tools are needed for certain jobs.

STRONG PEOPLE SKILLS

People skills are also called interpersonal skills. They are a huge part of a technician's job. Sometimes technicians must deal with frustrated clients. These clients may be working under tight deadlines. Technicians can help by staying composed while talking to the customer. It is important to show empathy. This can be done by understanding where the customer is coming from. Technicians can do this by listening carefully and not interrupting.

An IT support technician's tool kit may include screwdrivers, wire cutters, and cleaning tools.

Screwdrivers are used to help install or take apart hardware. Sometimes the technician must clean the device. Items such as compressed air canisters, microfiber cleaning cloths, liquid screen cleaners, and vacuums are helpful. It is also necessary to have various cords, chargers, and other items on hand. These tools help ensure everything is running properly when fixing computers, phones, and other equipment.

SAFETY AND RISK

Technicians usually don't have to worry about an unsafe working environment. Occasionally, they may need to work in

IT support technicians must exercise caution when connecting equipment in hard-to-reach places.

tight or hard-to-reach workspaces. This can happen if they have to crawl under a desk to plug in a computer or run some cords. It can be hard to move around in these areas.

Sometimes, furniture must be moved to reach electrical outlets or devices such as modems. Technicians may need to be strong enough to lift, push, or pull large objects such as desks and chairs. Technicians should be careful when doing these tasks. It is important to prevent injury or strain.

There are also possible hazards when working with electrical devices.

Frayed or damaged electrical wires are dangerous. They can cause electrical shock or fire.

Damaged outlets or frayed electrical cords and cables can cause problems. The technician has to look for signs that the equipment is not in good condition. They must not use it if it is dangerous.

CHAPTER FOUR

WHAT IS THE FUTURE FOR IT SUPPORT TECHNICIANS?

A career as an IT support technician can be a satisfying and important one. Help desk specialist Mai Yia Vue remarks on what she enjoys most about the job. "The best part of my day is when, as a Tier 1 specialist, I am able to resolve the

Helping others can be a rewarding part of an IT support technician's job.

issue. . . . It feels great to be able to help a colleague," she says. "I'm glad I'm part of this important team."[9] People who are interested in IT support technician jobs have a passion for computers and technology. They also love to help people.

BRIGHT OUTLOOK

In 2020 the US Bureau of Labor Statistics reported more than 844,000 IT support technicians working in the country. At that time, the need for IT support technicians was expected to continue to increase. This need was predicted to grow by about 9 percent between 2020 and 2030. That means about 70,000 new jobs opening up every year. Industries such as retail and health care are relying more on IT to serve clients. This causes the need for technicians to grow.

According to the BLS, the median pay rate for technicians is $55,510 per year. Some technicians make less than this. Others make more. Technicians who take on increased responsibility tend to have higher salaries. So do those who

WORKING DURING A PANDEMIC

The COVID-19 pandemic increased the need for technicians. More people started working from home instead of in an office. To work from home, people needed good computers and strong networks to run them. Technicians helped with setting up devices and fixing any problems that came up.

supervise others. A technician's pay will vary depending on the industry, company, and role.

DIVERSITY IN IT SUPPORT

Currently, more than 72 percent of people working in IT are white. As **entrepreneur** Tyler Bray remarks, there is huge value in having a more **diverse** workforce. He says, "Listen to new perspectives from people from all walks of life. It will only give you a broader mind and worldview, which supercharges your decision-making."[10] The IT industry can encourage more diversity by creating opportunities for those

RACE AND GENDER AMONG INDIVIDUALS IN IT POSITIONS

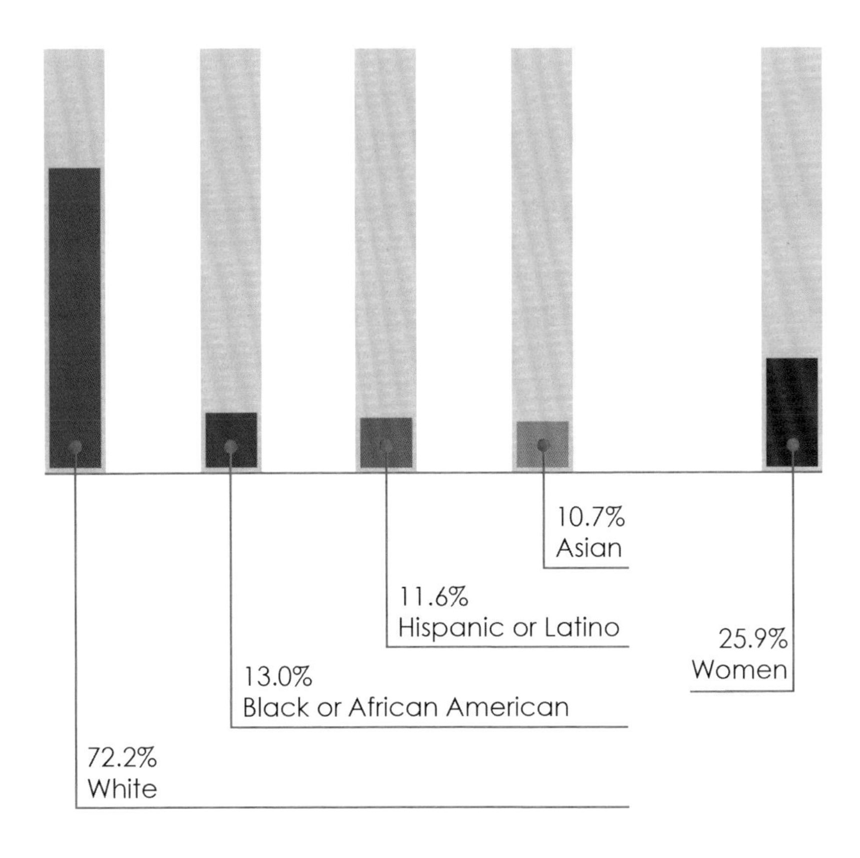

Source: "Labor Force Statistics from the Current Population Survey," US Bureau of Labor Statistics, *January 2021. www.bls.gov.*

In 2021 the Bureau of Labor Statistics studied the demographics of the IT support field. From a population of 660,000 computer support specialists, it found the above statistics about these workers.

workers to get better-paying, skilled jobs. They need to be able to feel welcome in the workplace. Hiring practices, training, and how the workplace is managed are all key for creating a comfortable place for everyone to work.

GENDER BIAS

Gender bias happens when one gender is preferred over another. It can happen in workplaces, including technology fields. More men are hired than women in the IT industry. But some changes are being made. Women-only tech groups are forming across the country. These serve as meeting

Women in tech workgroups can provide support, opportunities, and advice to other women in the male-dominated field.

places for women in tech. There, women can network and talk about their work. These groups can be support systems for

High school teachers can help students learn about opportunities in IT.

women in this largely male field. Women in these groups can provide resources and encourage each other.

Getting women interested in IT jobs is important. There are many ways to do this. One women-owned tech company matches high school teachers with IT professionals. The teachers hear what types of jobs are available in IT. They also learn how to train for them. The teachers then take that information back to the classrooms. They help their students learn about working in the IT field. This can help more girls learn about and want to work in IT.

IT support technicians work to keep the digital world running smoothly.

KEEPING THE WORLD PLUGGED IN

Technicians work to keep our devices, networks, and lives plugged in. Vue works in Minnesota for a large national bank. She provides technical support for the whole company. "My colleagues and I are the first line of defense for issues that happen within the technology side of the company," she says. "It's a huge responsibility, as issues can range from software features not working correctly to whole servers being down."[11]

DIGITAL HEALTH CARE

Technology is growing in the health care system. The ability for a patient and doctor to connect virtually is continuously increasing. Patients can schedule appointments and refill prescriptions. They can message doctors with their devices. They also may meet with their doctors over video call. Technicians play an important role in keeping these services running.

A career in IT support can be rewarding, interesting, and fulfilling.

Technicians often work behind the scenes. But their work can be seen all around. They keep the computers in homes, businesses, hospitals, government offices, and much more online. Their support helps day-to-day life run smoothly, both in and out of the office.

GLOSSARY

administrators

people in charge of a business or organization

apprenticeship

training on the job instead of in school

certifications

official documents earned for having certain qualifications or meeting certain standards

diagnose

to determine what causes a problem

diverse

people from a range of genders, sexual orientations, and social and ethnic backgrounds

entrepreneur

a businessperson

remotely

done from a distance

terminated

ended

SOURCE NOTES

INTRODUCTION: A DAY IN THE LIFE

1. "I.T. Support for School Districts - Day in the Life," *YouTube*, uploaded by I.T. Career Questions, May 29, 2019. www.youtube.com.

CHAPTER ONE: WHAT DOES AN IT SUPPORT TECHNICIAN DO?

2. Quoted in Emily Matzelle, "A Day in the Life of an IT Pro," *CompTIA*, May 19, 2017.

3. Quoted in "What Does an IT Technician Do?" *YouTube*, uploaded by Walser Automotive Group, April 22, 2021. www.youtube.com.

CHAPTER TWO: WHAT TRAINING DO IT SUPPORT TECHNICIANS NEED?

4. Quoted in "Build Your Future - IT Support Technician," *YouTube*, uploaded by Jackson Civils, July 25, 2017. www.youtube.com.

5. Quoted in "IT Jobs - Holland Hospital," *YouTube*, uploaded by Holland Hospital, August 31, 2015. www.youtube.com.

6. Quoted in "A Day in the Life of an IT Pro."

CHAPTER THREE: WHAT IS LIFE LIKE AS AN IT SUPPORT TECHNICIAN?

7. "A Day in the Life of a Help Desk Technician," *YouTube*, uploaded by IT Career Guide, January 1, 2021. www.youtube.com.

8. Quoted in "IT Computer Support Specialist," *YouTube*, uploaded by Western Technical College, March 2, 2012. www.youtube.com.

CHAPTER FOUR: WHAT IS THE FUTURE FOR IT SUPPORT TECHNICIANS?

9. Quoted in "A Day in the Life of an IT Pro."

10. "Eight Important Ways to Promote Inclusion and Diversity in Your Workplace," *Forbes*, January 25, 2021. www.forbes.com

11. Quoted in "A Day in the Life of an IT Pro."

FOR FURTHER RESEARCH

BOOKS

Leanne Currie-McGhee, *Cutting Edge Careers in Info Tech.* San Diego, CA: ReferencePoint Press, 2021.

Stuart A. Kallen, *Exploring Hi-Tech Careers.* San Diego, CA: ReferencePoint Press, 2021.

Helen Mason, *Cutting-Edge Careers in Technical Education: Dream Jobs in Information Technology.* Ontario, Canada: Crabtree, 2018.

INTERNET SOURCES

Alison Doyle, "Important Skills for Technical Support Jobs," *Balance Careers,* April 2020. www.thebalancecareers.com.

"Information Technology Support Technician Job Description," *Job Hero,* September 2021. www.jobhero.com.

"What Is an IT Technician? How to Become One," *Coursera,* October 2021. www.coursera.org.

WEBSITES

Career Outlook

www.bls.gov/careeroutlook/home.htm

Career Outlook provides data and information about occupations and industries, including topics such as pay and training.

CompTIA

www.comptia.org/membership/it-pro

CompTIA IT Pro and Student Membership (formerly the Association of Information Technology Professionals) focuses on education and learning opportunities for people employed in IT.

HelpDesk Chapters

www.hdilocalchapters.org/cpages/home

HelpDesk Chapters is a member-based organization that helps IT support workers connect and network.

INDEX

IMAGE CREDITS

Cover: © Iryna Rahalskaya/ Shutterstock Images
5: © Bojan Milinkov/ Shutterstock Images
7: © Tyler Olson/Shutterstock Images
8: © fizkes/Shutterstock Images
11: © Eimantas Buzas/ Shutterstock Images
12: © ESB Professional/ Shutterstock Images
15: © 2p2play/Shutterstock Images
17: © Hand Robot/ Shutterstock Images
19: © Africa Studio/ Shutterstock Images
20: © fizkes/Shutterstock Images
22: © A_stockphoto/ Shutterstock Images
25: © dotshock/Shutterstock Images
29: © Gorodenkoff/ Shutterstock Images
31: © ESB Professional/ Shutterstock Images
34: © dotshock/Shutterstock Images
36: © Monkey Business Images/ Shutterstock Images
39: © Axtem/Shutterstock Images
41: © Monkey Business Images/ Shutterstock Images
42: © fizkes/Shutterstock Images
45: © Phovoir/Shutterstock Images
49: © Tyler Olson/Shutterstock Images
51: © Song_about_summer/ Shutterstock Images
54: © BublikHaus/Shutterstock Images
56: © Alpa Prod/Shutterstock Images
58: © Armands Photography/ Shutterstock Images
61: © fizkes/Shutterstock Images
65: © Red Line Editorial
67: © Jacob Lund/ Shutterstock Images
68: © Jacob Lund/ Shutterstock Images
70: © Tyler Olson/Shutterstock Images
72: © fizkes/Shutterstock Images

ABOUT THE AUTHOR

Sheryl Normandeau is a Calgary-based writer. Her nonfiction and fiction work has been published internationally.